AF207092

MERCURY

by Carolyn Bennett Fraiser

BrightPoint Press

San Diego, CA

© 2026 BrightPoint Press
an imprint of ReferencePoint Press, Inc.
Printed in the United States

For more information, contact:
BrightPoint Press
PO Box 27779
San Diego, CA 92198
www.BrightPointPress.com

LIBRARY OF CONGRESS CATALOGING-IN-PUBLICATION DATA

Name: Fraiser, Carolyn Bennett, author.
Title: Mercury / by Carolyn Bennett Fraiser.
Description: San Diego, CA: ReferencePoint Press, 2026 | Series: The planets of our solar system | Audience: Grade 7 to 9 | Includes bibliographical references and index.
Identifiers: ISBN: 9781678211660 (hardcover) | ISBN: 9781678211677 (eBook)
The complete Library of Congress record is available at www.loc.gov.

CONTENTS

AT A GLANCE

- Mercury is the smallest, fastest planet in the solar system. It is also the closest planet to the Sun.

- Mercury orbits the Sun every 88 Earth days. It rotates only three times every 2 Mercury years.

- Temperatures on Mercury swing from −290°F (−180°C) to 800°F (430°C). The Sun's heat quickly escapes into space.

- Mercury has no moons. This is because the Sun's strong gravity pulls any objects away.

- Mercury has been studied since ancient times. Early astronomers tracked its movement in the sky.

- Telescopes helped astronomers create the first maps of Mercury's surface.

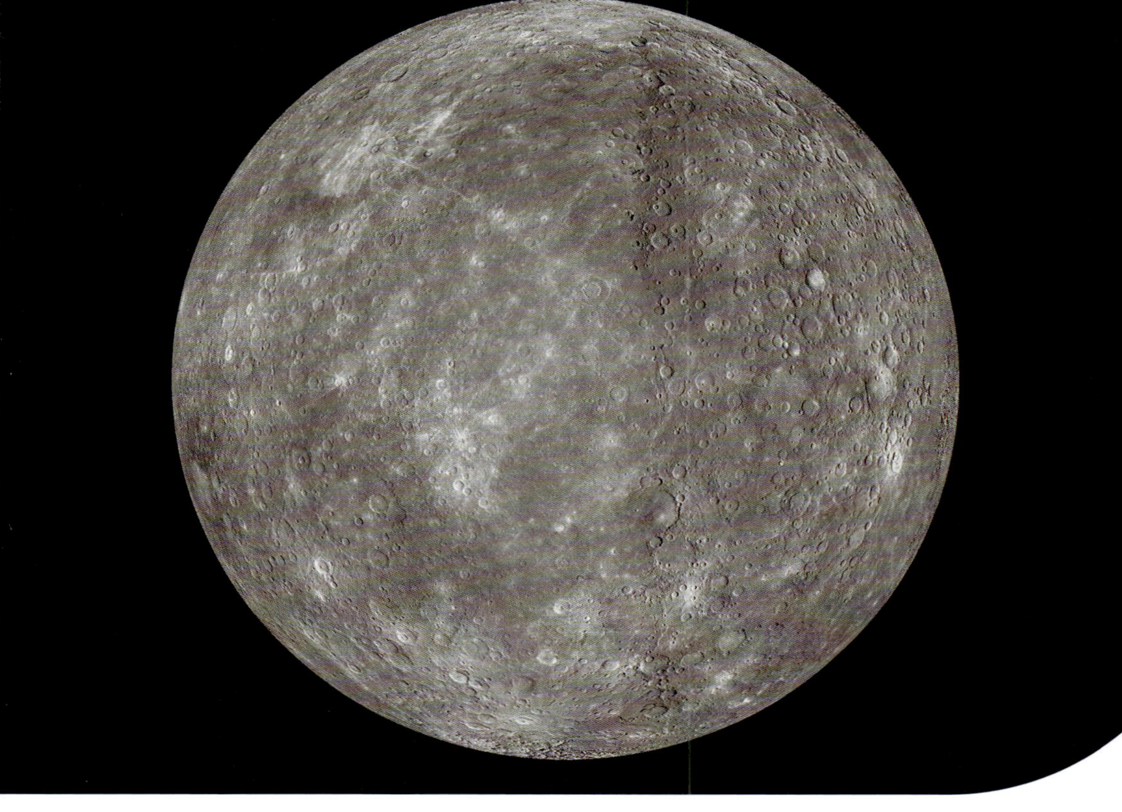

- Mercury is one of the most difficult planets to reach. Missions have to battle the Sun's strong heat and gravity to enter orbit.

- Only three missions have sent spacecraft to Mercury. Two of these spacecraft are part of the BepiColombo mission.

- Mercury has many surface features that interest scientists. These include craters and long, tall scarps.

A SURPRISING DISCOVERY

In 1991, scientists pointed a large dish antenna toward the inner solar system. They hoped to learn more about one of the least-known planets, Mercury. This burning-hot planet is the closest planet to the Sun. Sending spacecraft to it is difficult. Only one **probe** had flown by Mercury in the 1970s. It had mapped less than half the surface. Much remained unknown about

Dish antennas work by reflecting beams from space onto a single point in the center of the dish. Devices located at this point collect the beams and record the information they carry.

MESSENGER could capture Mercury's surface in much greater detail than ground-based technology.

the planet. The scientists hoped to map Mercury's unexplored side.

The antenna sent beams to Mercury's surface. The scientists waited for the beams to bounce back to Earth. Large telescopes read the reflections as various colors. Blue stood for the darkest areas. Red represented the brightest spots. The colors would help them map the planet's surface. They were shocked by what they saw.

The reflections from Mercury's north and south poles were red. This could mean many things. But scientists were excited about one possibility: ice. Similar spots had been seen on Mars and some of Jupiter's moons. But few expected to find ice on one of the hottest planets.

Scientists would have to wait until 2012 to confirm the presence of ice. That was when the next probe visited Mercury. The probe was called *MESSENGER*. It was launched by the National Aeronautics and Space Administration (NASA). And it was well worth the wait. The probe found ice on Mercury's north pole. The ice was hiding deep within craters that never saw sunlight. Shadows shielded the ice from the Sun's heat.

Parts of this crater near Mercury's north pole are in permanent darkness.

STUDYING MERCURY

Finding ice was just the beginning. Today scientists are learning more about the smallest planet. They think that Mercury might hold clues about how the solar system began.

Scientist Johannes Benkhoff is excited to learn more. He works with the latest mission to Mercury, BepiColombo. This mission launched spacecraft toward Mercury in 2018. He says, "Fifteen years ago, Mercury was considered a boring planet. But I expect to find many more surprises."[1]

CHAPTER ONE

MEET MERCURY

Mercury is the closest planet to the Sun. Like Venus, Earth, and Mars, it is a terrestrial planet. It has a rocky surface and inner core. Scientists believe Mercury is one of the oldest planets. They think it was once as large as Earth. Their theory is that it collided with another planet. The impact blasted away some of its mass.

Today, Mercury is about one-third the diameter of Earth. It is the smallest planet in

The *MESSENGER* spacecraft took this image of the shadowy line separating Mercury's light and dark sides. This line is called a terminator.

the solar system. Ancient lava left its surface very dark. The hardened lava is marked with tens of thousands of craters. The craters came from meteors and comets that crashed into the planet.

As the closest planet to the Sun, Mercury orbits the fastest. This means it has the shortest year. Mercury zips around the Sun at about 107,082 miles per hour

(172,332 kmh) on average. At that speed, it takes only 88 Earth days to complete one orbit.

Mercury's orbit is unusual. Most planets have orbits that are nearly circular. But Mercury's is more egg-shaped. At its most distant point, Mercury is 1.5 times farther from the Sun than at its nearest. The Sun's gravity causes Mercury to move faster as it

No Moons for Mercury

Mercury and Venus are the only moonless planets. Mercury's gravity is too weak compared to the Sun's to hold onto a moon. Mercury's gravity is only one-third as strong as Earth's. The Sun's gravity is twenty-eight times stronger than Earth's. The closeness of the Sun to Mercury means it would easily pull a moon away from the planet.

nears the Sun. Then the planet slows as it moves away.

At first glance, Mercury resembles Earth's Moon. They are similarly sized. Both have ancient crater impacts. And both lack a substantial **atmosphere**. But the two are very different in other ways. Scientists have learned quite a bit about the Moon. But they still have a lot of questions about Mercury. Scientist David Rothery says, "This is a planet that we really do not yet fully understand."[2]

CLOSEST BUT NOT THE HOTTEST

Temperatures on Mercury can soar as high as 800°F (430°C). That's hot enough to melt lead. The Sun appears three times as large in Mercury's sky as in Earth's. And its light

One difference between the Moon and Mercury is that Mercury's sunlit side can get more than 500°F (260°C) hotter than the Moon's.

is seven times brighter. But Mercury is not the hottest planet in the solar system. Venus is. Venus's thick atmosphere traps heat like a blanket. Mercury doesn't have a thick atmosphere. Instead, it has an exosphere. This layer of gas is very thin. It is too thin to hold much heat. Warmth quickly escapes

Mercury's lack of a substantial atmosphere means that it does not have clouds.

into space. Temperatures drop to –290°F (–180°C) in the shade. Mercury's natural temperature range is the widest in the solar system.

Mercury's exosphere is one-trillionth as dense as Earth's atmosphere. It contains oxygen. But there is not enough for humans to breathe. Mercury's exosphere also has sodium, hydrogen, helium, and potassium **atoms**. These small particles rarely bump into each other. They fall to the surface and bounce up like rubber balls.

Mercury turns very slowly. It rotates once every 58 Earth days. That means it only rotates three times every 2 Mercury years. This creates unusual sunrises and sunsets. Areas at the poles stay in total darkness. Regions near the equator have

constant sunlight. In some places, the Sun
does not move in the sky for weeks. In
other spots, it rises and sets several times
in a single day.

Mercury's seasons are also unusual.
Earth's tilt on its axis creates its seasons.
But Mercury sits almost upright. And
because the exosphere is so thin, there
is no weather. Instead, Mercury heats up
when it nears the Sun. It cools down as it
moves farther away.

MERCURY'S CORE

The cores of most terrestrial planets are
about one-third their size. The rest is rock.
But Mercury has a huge iron core. It takes
up about 75 to 85 percent of its total size.
That leaves a thin, rocky crust.

MERCURY'S INNER STRUCTURE

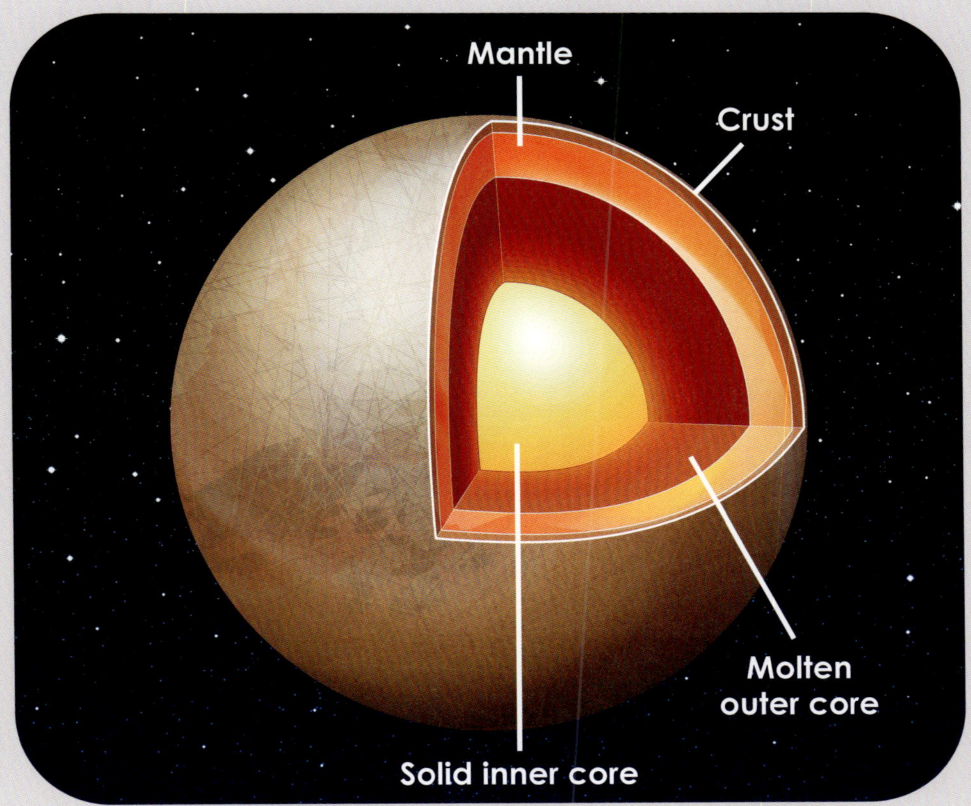

This diagram displays the layers beneath Mercury's surface. The mantle is a layer of partially molten rock.

At first, scientists thought Mercury's entire core was solid. They assumed that it had cooled over time. But evidence later showed that the outer part of Mercury's core is molten. They think that this part of

the core may contain something other than iron. A lighter element such as sulfur might cool more slowly. This would allow it to stay molten for longer.

Mercury also has a **magnetic field**. It is the only inner planet besides Earth to have one. But Mercury's field is one hundred times weaker than Earth's. Over time, Mercury's core will cool. This may affect the planet's magnetic field. Scientist Antonio Genova says, "Mercury may help us predict how Earth's magnetic field will change as the core cools."[3]

STUDYING MERCURY FROM EARTH

Mercury can easily be seen from Earth. Stargazers do not need a telescope or binoculars. The planet always appears close to the Sun. It is best seen shortly before sunrise or after sunset.

Before the invention of spaceflight, **astronomers** watched Mercury from Earth. The planet was visible but hard to study. It was much smaller than other planets. And it could only be seen when it

Some of the earliest recorded observations of Mercury are found in clay tablets made by Babylonians.

was closest to the Sun. But the Sun makes observing Mercury from Earth difficult. Its glare blocks views of the planet's surface. And its brightness can damage telescopes that are not designed to be pointed at the Sun.

FIRST EYES ON MERCURY

No one knows who first discovered Mercury. Humans have watched it since ancient times. The Babylonians wrote about five stars in the 700s BCE. They looked to the stars to predict the future. Each was named after a god. Mercury was called the Star of Nabu.

But the Babylonians were not the only ones gazing at Mercury. In China, the planet was associated with the element of water.

To the Egyptians, it was the god Thoth. Norse cultures called it Odin after one of their gods. The Maya tracked its location in the sky. A Greek astronomer created a star chart. It included Mercury and the four other planets. He recorded their positions each day.

Watching the Skies Without Telescopes

Before telescopes, astronomers studied the skies with the naked eye. Many used other inventions. Quadrants measured the stars by their height above the horizon. Astrolabes used the position of the stars to calculate time and dates. Both helped scientists track movement and patterns. Scientists even used these devices to make predictions about future events in the night sky.

The Greeks continued to watch the night sky. They looked for patterns. They noticed that five bright objects in the night sky moved differently than most stars.

The Greeks called these objects planets, or wandering stars. Early scientists recorded their orbits in diaries. In the 100s BCE, Egyptian astronomer Ptolemy watched Mercury and the other planets. He believed the planets circled Earth.

The Greeks named the planets after their gods. At first, they thought Mercury was two different planets. They called it Apollo in the morning and Hermes in the evening. Then they realized it was the same planet. They kept the name Hermes, after their messenger god. Later, the Romans used the name Mercury. Mercury is a Roman version of the Greek god Hermes. The other visible planets were named Venus, Mars, Jupiter, and Saturn. These names remain in use in English and many other languages today.

THE INVENTION THAT CHANGED EVERYTHING

People did not learn much more about Mercury until the mid-1500s. That is when Nicolaus Copernicus introduced a new idea. He suggested that the planets orbit around the Sun, not Earth. Scientists were divided. Some sided with Copernicus. But others stuck with Ptolemy's view of Earth being at the center.

In the early 1600s, people invented the telescope. Italian astronomer Galileo Galilei was one of the first to point a telescope at the planets. He noted that the planets looked like little moons.

They were spherical. A few years later, scientist Pierre Gassendi noticed the same thing. He used a special device to watch Mercury pass in front of the Sun. The planet appeared as a black dot against the Sun. Its journey took 5 hours. At first, Gassendi thought the black dot was something else. He wrote, "I was far from suspecting that Mercury would project

Nicolaus Copernicus was born in Poland. This statue of the astronomer stands in Toruń, the city where he was born.

such a small shadow."[4] It was half the size he expected. His report forced scientists to rethink the size of the known planets.

The new observations of Mercury confirmed Copernicus's idea. Planets orbit the Sun, not Earth. People also learned that planets are closer to Earth than the stars. And they learned that the planets do not move in perfect circles.

The Sun made studying Mercury's surface challenging. Astronomers drew maps of what they could see. They noted dark patches on its surface. They did not see any signs of an atmosphere. They believed the Sun had scorched Mercury. It was seen as a dead world with no water or weather. Anything else about Mercury was a mystery.

CHAPTER THREE

EXPLORING MERCURY

Scientists want to learn more about Mercury. Planetary scientist Johannes Benhkoff wonders if life could be found there. He says, "You have temperatures of 450°F [(230°C)] like a pizza oven. Then you have water ice in the craters . . . . Imagine if we were the first to find [signs of life] in these craters."[5]

But getting to Mercury is not easy. Probes have to endure extreme heat

Spacecraft need a lot of fuel to reach Mercury. More than half of the *MESSENGER* spacecraft's mass was fuel.

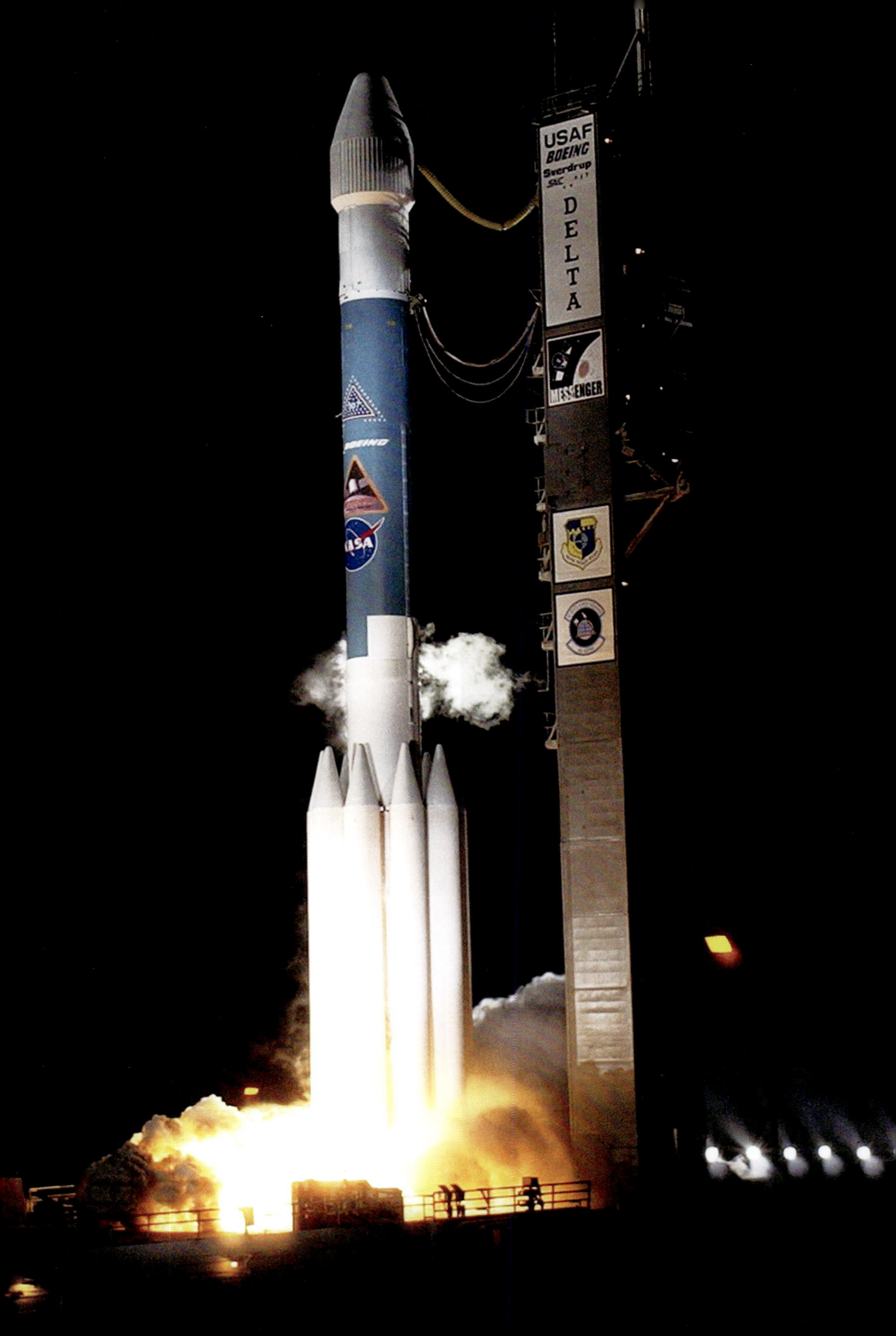

USAF
BOEING
Sverdrup
DELTA
MESSENGER

This backup version of *Mariner 10* is on display at the National Air and Space Museum in Washington, DC.

and cold. And they have to fight against the Sun's gravity. That requires a lot of fuel. It takes more fuel to reach Mercury than to leave the solar system.

MARINER 10

Despite the challenges of visiting Mercury, NASA scientists found a way. In 1973, they launched *Mariner 10*. It became the

first probe to fly by Mercury. A sunshade
shielded the spacecraft from the Sun's heat.
Mariner 10 was the first spacecraft to use
a new method to save fuel. This method is
called a gravity assist. The probe flew by
Venus. Then it used the planet's gravity to
sling itself toward Mercury.

On March 16, 1974, *Mariner 10* flew by
Mercury. The probe's cameras took photos
of the mysterious planet. For the first time,
scientists saw its surface in detail. Large
scarps stretched for hundreds of miles.
Countless craters marked the surface. One
was almost 1,000 miles (1,600 km) across.
Mariner 10 also proved that Mercury
had a thin atmosphere. And it made an
unexpected discovery. The planet had a
magnetic field.

Scientists were surprised. A magnetic field indicated that Mercury had a liquid core. They had once thought the planet was too small to have a liquid core. Instead, its core should have been solid. Mercury only has one **tectonic plate**. A liquid core meant the plate was still sliding across Mercury's surface.

After flying by Mercury once, *Mariner 10* circled the Sun. It returned to Mercury two more times. The final flyby was the closest. It passed only about 200 miles (320 km) above the surface. But a problem with the probe's data systems occurred. Only a few photos from that pass made it back to Earth. By 1975, the probe's fuel was running out. It had taken more than 2,700 photos. Scientists used them to map 45 percent of

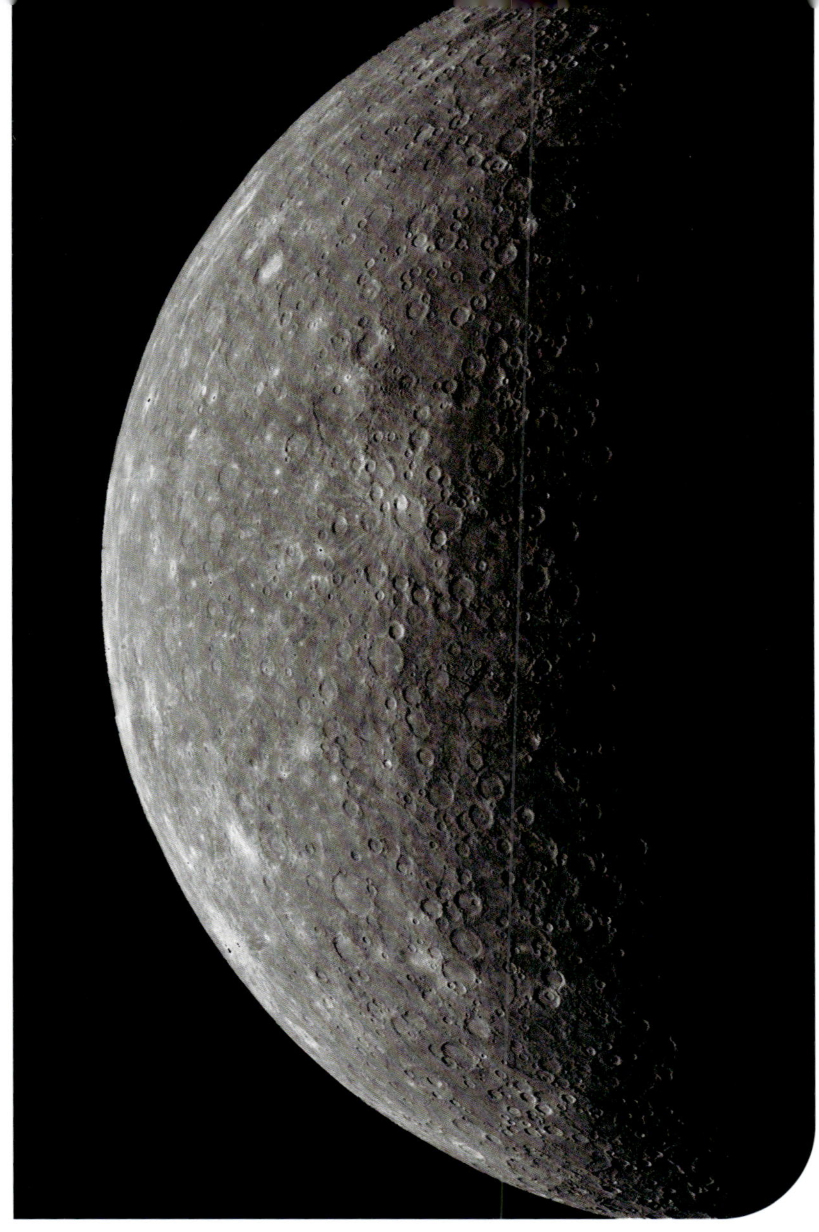

This is the first picture that *Mariner 10* took of Mercury.

Mercury's surface. On March 24, NASA lost contact with *Mariner 10*. The spacecraft is probably still in orbit around the Sun.

Scientists had more questions than answers. They wanted to learn more about Mercury. But they would have to wait more than 30 years for their next opportunity to visit.

MESSENGER

In 2004, NASA was ready to return. On August 3, *MESSENGER* blasted off into space. It was the second mission to Mercury. *MESSENGER* circled the Sun fifteen times. Then, in 2011, it made history. It entered orbit around Mercury.

For the first time, scientists saw Mercury in color. *MESSENGER* took photos using camera filters. The filters picked up the colors of many minerals. The photos identified craters of various ages and sizes.

This NASA illustration shows *MESSENGER* orbiting Mercury.

They also captured many more scarps. Scientists have learned that these features formed as the planet slowly shrank over billions of years. This shrinking occurs as Mercury's interior cools. As the planet shrinks, its surface cracks.

MESSENGER also revealed something new about Mercury. Shallow pits appear on the surface. But they are not shaped like craters. They do not have rims. Scientists believe they are volcanic vents. Lava once blew through these weak spots in Mercury's crust. That lava may have covered the

MESSENGER discovered and photographed this large basin created by an impact on Mercury's surface.

surface in a layer 3 miles (5 km) deep. The lava flooded craters. And it created large, smooth fields.

MESSENGER circled Mercury for more than 4 years. It mapped 98 percent of the planet. Scientists studied Mercury's surface, core, and magnetic field. The mission also confirmed what they had been waiting for. Ice did exist at the poles.

By the end of 2014, *MESSENGER* was close to running out of fuel. Scientists predicted that the spacecraft would soon crash into Mercury. This happened on April 30, 2015. The crash formed a crater.

BEPICOLOMBO

Scientists did not wait long for the next mission to Mercury. Space programs in

Europe and Japan teamed up to launch BepiColombo. This mission involves two probes studying Mercury at the same time. They blasted off on the same rocket on October 20, 2018. The probes were designed to separate once in orbit around Mercury.

Scientists planned for the probes to fly by Mercury six times. The first flyby happened in October 2021. The sixth

Giuseppe Colombo

The BepiColombo project was named after Giuseppe "Bepi" Colombo. Colombo was born in Italy in 1920. He was a mathematician, scientist, and professor. He advised NASA to use a gravity assist to reach Mercury. This advice was key to the success of *Mariner 10*.

happened in January 2025. The probes were originally expected to enter Mercury's orbit in late 2025. But an issue with their rockets led to a delay. The rockets would not be able to work at full power. Mission scientists were able to chart a new course for the spacecraft that required less power. The plan was for BepiColombo to enter orbit in late 2026.

After separating, the two spacecraft would circle the planet at the same time. But they would fly over different areas. One would collect data about Mercury's surface and core. The other would focus on the planet's magnetic field. Together, scientists expected the probes to reveal more details about Mercury.

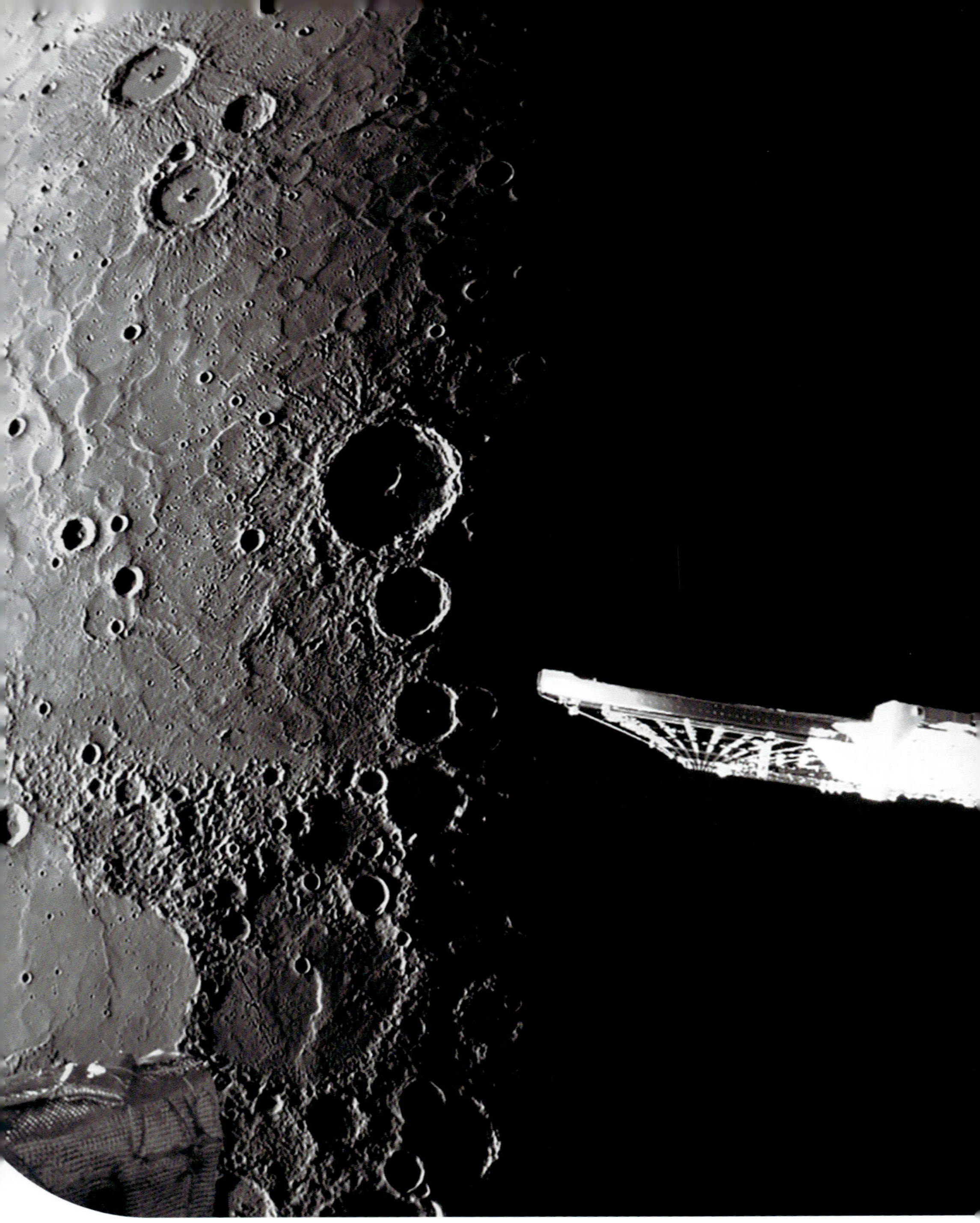

BepiColombo took this picture of itself during one of its flybys over Mercury's surface.

The mission originally had a third part. Scientists wanted to put a lander on Mercury's surface. But the cost was too high. This part was canceled. Scientists have proposed a new mission that would launch in 2035. It would land a probe on Mercury's surface, helping to reveal more about the mysterious planet.

CHAPTER FOUR

A PLANET LIKE NO OTHER

Mercury is unique in many ways. As its core cooled, the crust shrunk. Its tectonic plate cracked and bent. This created giant scarps. But Mercury also has younger, smaller scarps. They would not have survived ancient meteor strikes.

These young scarps excite scientists like Thomas Watters. He says, "[This] means that Mercury joins Earth as a tectonically active planet."[6] Scientists believe Mercury's

Cracks cross Mercury's surface in this picture taken by *MESSENGER*.

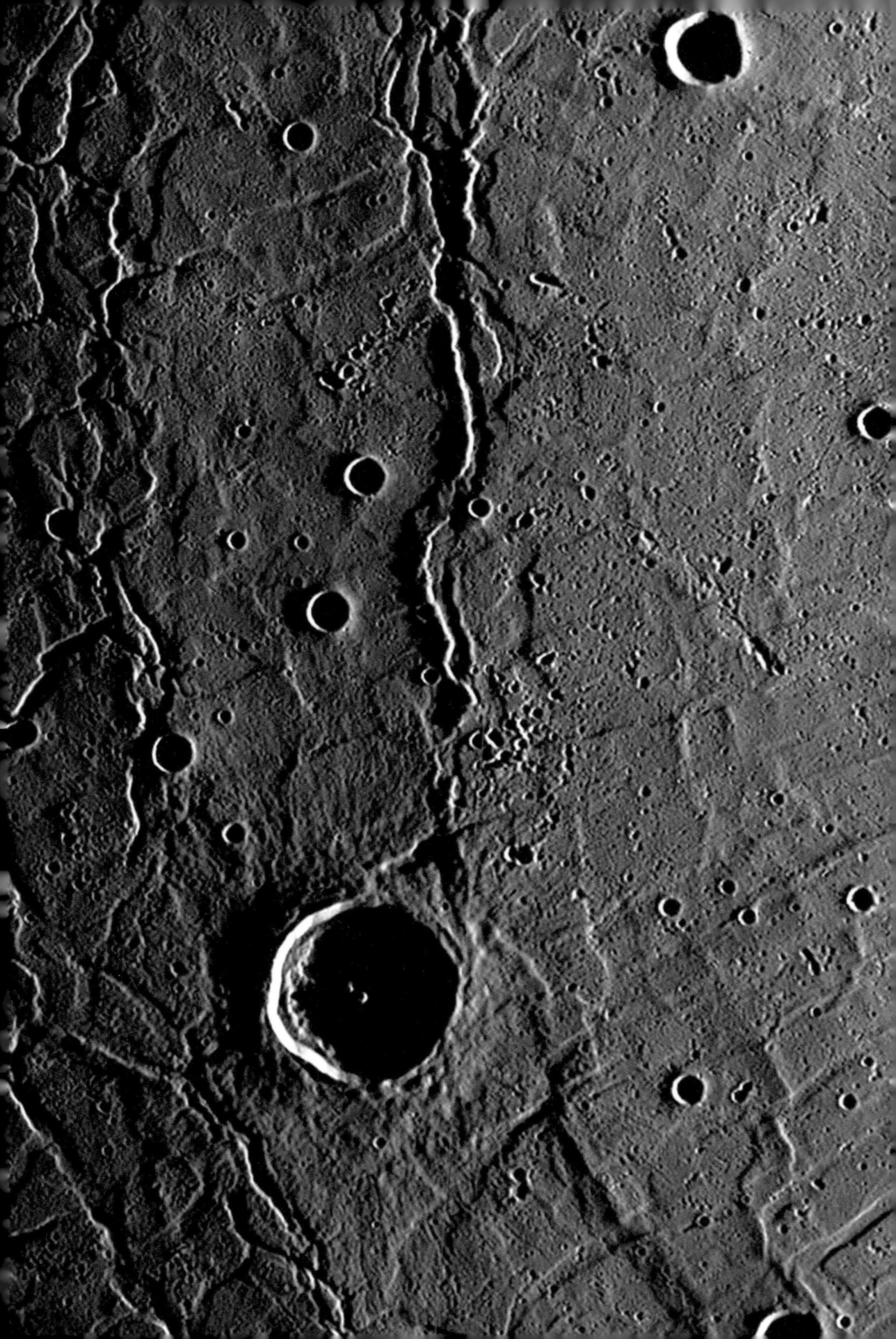

core has somehow kept some heat. It could still be active. Or it may have been active recently.

As the planet shifted, pieces of Mercury's crust bent upward. But other areas sank. This created large, smooth plains. At least 40 percent of Mercury's surface is flat. Wide, smooth valleys stretch between the

This smooth plain covers part of Mercury's northern region.

high cliffs. Tectonic activity may have been involved. Mercury's Great Valley is more than 620 miles (1,000 km) long. Two giant scarps are on either side. Some rise several miles high. Mercury's surface features can provide clues about its core.

THE CALORIS BASIN

The Caloris Basin is one of Mercury's largest craters. It measures about 960 miles

The Beagle Rupes

Mercury has some of the largest scarps in the solar system. One of the largest cliffs on Mercury is the Beagle Rupes. This long, curved cliff stretches for more than 372 miles (599 km). It was named after a ship that carried British biologist Charles Darwin.

(1,550 km) across. It is also one of the oldest. Scientists believe it formed 3.8 billion years ago. An **asteroid** impact most likely created it. Inside are several smaller, younger craters.

The Caloris Basin is not an ordinary crater. Near the center are 200 long, narrow pits. They create a weblike pattern around a small crater. Scientists call this feature Pantheon Fossae. The Pantheon is a domed building in Rome. *Fossae* is Latin for *trenches*. Scientists have not found this pattern anywhere else on Mercury. They are not sure how it formed.

Scientists are interested in this crater for many reasons. *MESSENGER* found volcanic vents around its edges. Smaller craters on the inside are also mysterious. They have

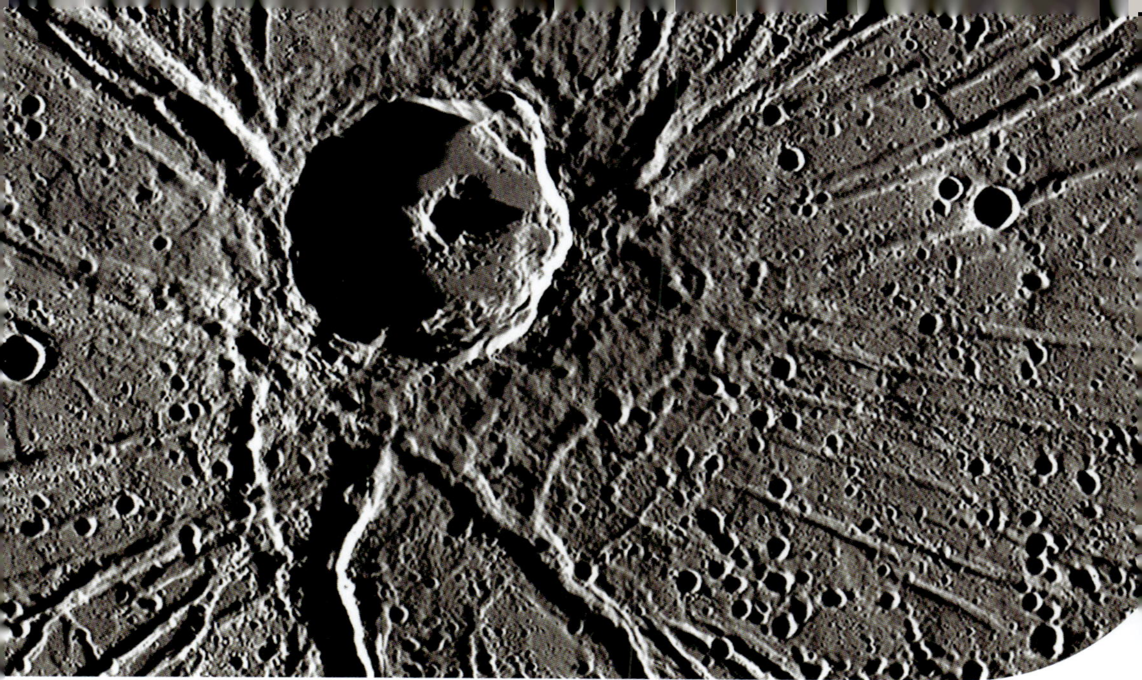

unusual dark and light spots. They could reveal more about Mercury's history. Astronomer Frank Summers says, "Mercury lacks wind, rain, volcanoes, or other erosion that would erase the craters."[7]

MERCURY'S LONG TAIL

Mercury leaves behind a stream of material as it moves through space. This tail is a result of the Sun. Strong **solar wind** blows sodium and other elements from Mercury's exosphere into space. The tail stretches for about 15 million miles (24 million km) on average. It gets longer as Mercury nears the Sun. It shrinks as the planet moves away. Sunlight gives it a yellow glow.

Scientists in the 1980s first suspected Mercury might have a tail. But it wasn't

Tails are more commonly observed trailing comets. Comets are icy, rocky objects that shed some of their material behind them due to the Sun's heat.

spotted until 2001. Then, in 2008, *MESSENGER* flew through the tail. The probe discovered calcium and magnesium in it. Scientists think they can learn more from Mercury's tail. They want to understand how the Sun affects planets' magnetic fields.

People can see Mercury's tail from Earth. But sky gazers need to use a camera or telescope. It is best seen when Mercury is brightest. This happens when the planet is closest to the Sun.

MYSTERIOUS HOLLOWS

Mercury holds even more mysteries. Low regions called hollows surrounded by bright bluish patches appear on its surface. These hollows are very shallow. They are various shapes and sizes. Some are only a few feet wide. Others stretch for more than a mile (1.6 km). Inside, their floors are smooth. A bright ring or halo surrounds each hollow. The colors seem to darken as they age.

Scientists have not seen these hollows anywhere else in the solar system. They do not yet know how they formed. The Sun's heat might have blown off small pieces of rock. Unlike volcanic vents, they are not billions of years old. They are much younger. Scientists believe they formed in the last 100,000 years. And Mercury may still be creating them today.

These discoveries on Mercury excite scientists like Deborah Domingue. She says, "We expected this thing to have been beaten up, processed. It's done for.

It's baked and cooked. There's all this intriguing evidence that Mercury is not this . . . piece of stone."[8]

BepiColombo and future missions will reveal more. Scientists are just beginning to uncover Mercury's secrets. Scientists hope these missions will answer some of the many questions about the mysterious world.

GLOSSARY

asteroid
a large rock moving through space

astronomers
scientists who study space

atmosphere
gases that surround a planet

atoms
basic building blocks of matter

magnetic field
the area around a magnetic material within which the force of magnetism works

probe
a device sent to space that gathers information

scarps
long, high cliffs on a planet's surface

solar wind
charged atoms that stream out from the Sun

tectonic plate
a large piece of solid rock composing part of a planet's crust

SOURCE NOTES

INTRODUCTION: A SURPRISING DISCOVERY

1. Quoted in Alessia Franco and David Robson, "Mercury: The Solar System's Smallest Planet May Once Have Been as Large as Earth," *BBC*, April 14, 2024. www.bbc.com.

CHAPTER ONE: MEET MERCURY

2. Quoted in "BepiColombo's First Views of Mercury," *The European Space Agency*, February 10, 2021. www.esa.int.

3. Quoted in Karl B. Hille, "A Closer Look at Mercury's Spin and Gravity Reveals the Planet's Inner Solid Core," *NASA*, April 17, 2019. www.nasa.gov.

CHAPTER TWO: STUDYING MERCURY FROM EARTH

4. Quoted in Albert van Helden, "The Importance of the Transit of Mercury of 1631," *Journal for the History of Astronomy*, 1976, pp. 1–10.

CHAPTER THREE: EXPLORING MERCURY

5. Quoted in Shannon Stirone, "Spacecraft Launching This Week Will Explore the Mysteries of Mercury," *Smithsonian Magazine*, October 17, 2018. www.smithsonianmag.com.

CHAPTER FOUR: A PLANET LIKE NO OTHER

6. Quoted in Alison Wood, "Mercury Joins Earth as Tectonically Active Planet," *Smithsonian*, September 26, 2016. www.si.edu.

7. Quoted in "Mercury," *American Museum of Natural History*, n.d. www.amnh.org.

8. Quoted in Shi En Kim, "The Seven Most Amazing Discoveries We've Made by Exploring Mercury," *Smithsonian Magazine*, April 16, 2024. www.smithsonianmag.com.

FOR FURTHER RESEARCH

BOOKS

Jane Parks Gardner, *Venus*. BrightPoint Press, 2026.

Gail Radley, *The Space Encyclopedia*. Abdo Publishing, 2023.

Meg Thacher, *The Terrestrial Planets: Mercury, Venus, Earth, and Mars*. BrightPoint Press, 2022.

INTERNET SOURCES

"BepiColombo," *NASA*, February 26, 2025. https://science.nasa.gov.

"Mercury 101," *National Geographic*, December 3, 2024. https://education.nationalgeographic.org.

"Mercury, World of Extremes," *The Planetary Society*, n.d. www.planetary.org.

NASA: Eyes on the Solar System
https://eyes.nasa.gov

On NASA's Eyes on the Solar System website, visitors can explore a simulation of the solar system. The simulation includes the positions of spacecraft associated with active NASA missions.

NASA: Mercury
https://science.nasa.gov/mercury

This official NASA website summarizes what people have learned about Mercury. The website provides many amazing pictures of the closest planet to the Sun.

The Schools' Observatory
www.schoolsobservatory.org

The Schools' Observatory guides visitors on how to observe the night sky. The website also has information about career paths in astronomy.

INDEX

IMAGE CREDITS

Cover: © NASA

5: © NASA

7: © NASA

8: © NASA

10: © NASA

13: © NASA

14: © NASA

17: © NASA

18: © NASA

21 (Mercury): © Designua/Shutterstock Images

21 (background): © Valery Brozhinsky/Shutterstock Images

22: © Reto Stöckli/NASA

25: © Hendrik Schmidt/dpa-Zentralbild/dpa picture alliance/Alamy

28: © NASA

31: © Genna Duberstein/NASA

32: © Kamil Saks/Shutterstock Images

35: © NASA

36: © National Air and Space Museum

39: © NASA

41: © NASA

42: © NASA

46: © European Space Agency/AP Images

49: © NASA

50: © NASA

53: © NASA

54: © Aaron Kingery/NASA

57: © NASA

ABOUT THE AUTHOR

Carolyn Bennett Fraiser grew up south of Kennedy Space Center in Florida and has loved space since she was a child. A former journalist, she loves to write about space, nature, and stories buried in history. She is the author of several other educational books for students. Carolyn currently lives in Brevard, North Carolina, where she helps kids and teens discover a passion for reading and writing.